Life Beyond Earth

Cameron Macintosh

Life Beyond Earth

Text: Cameron Macintosh
Publishers: Tania Mazzeo and Eliza Webb
Series consultant: Amanda Sutera
Hands on Heads Consulting
Editor: Kirstie Innes-Will
Project editors: Annabel Smith
Designer: Leigh Ashforth
Project designer: Danielle Maccarone
Permissions researchers: Lumina Datamatics
Production controller: Renee Tome

Acknowledgements
We would like to thank the following for permission to reproduce copyright material:

Front cover, p. 5 (top): Haitong Yu/Moment/Getty Images; p. 4: NASA/CXC/JPL-Caltech/STScl; p. 5: (bottom) Dmitriy Moroz/Alamy Stock Photo; content pages; back cover; p. 6: NASA Goddard/Reto Stöckli; p. 7: Science Photo Library/Alamy Stock Photo; p. 8: PandorumBS/Alamy Stock Photo; p. 9: Jet Propulsion Laboratory/NASA Images; p. 10: (left) MICHAEL J DALY/SCIENCE PHOTO LIBRARY; (middle) Science Photo Library/Alamy Stock Photo; (right) DENNIS KUNKEL MICROSCOPY/SCIENCE PHOTO LIBRARY; p. 11: Stocktrek Images, Inc./Alamy Stock Photo; p. 12: (left) Ulrich Doering/Alamy Stock Photo; (top) davidf/iStock/Getty Images; (middle) Hansen.matthew.d/Shutterstock.com; (right) xfox01/Shutterstock.com; p. 13: Signal Photos/Alamy Stock Photo; p. 14: Gremlin/E+/Getty Images; p. 15: Artur Maltsau/Alamy Stock Photo; p. 16: Kiyoshi Takahase Segundo/Alamy Stock Photo; p. 17: Win McNamee/Getty Images News/Getty Images; p. 18: NASA Image Collection/Alamy Stock Photo; p. 19: (top) NASA Images; (bottom) NASA/JPL; p. 20: Science History Images/Alamy Stock Photo; p. 21: (top) NASA/JPL-Caltech; p. 22: M2 Photography/Alamy Stock Photo; p. 23: matrioshka/Shutterstock.com; p. 24: New Africa/Shutterstock.com; p. 25: Universal History Archive/Universal Images Group/Getty Images; p. 26: Universal History Archive/Universal Images Group/Getty Images; p. 27: Chronicle/Alamy Stock Photo; p. 28: geogphotos/Alamy Stock Photo; p. 29: Decosu~commonswiki; p. 30: NASA.

Every effort has been made to trace and acknowledge copyright. However, if any infringement has occurred, the publishers tender their apologies and invite the copyright holders to contact them.

NovaStar

ISBN 978 0 17 033522 5

Cengage Learning Australia
Level 5, 80 Dorcas Street
Southbank VIC 3006 Australia
Phone: 1300 790 853
Email: aust.nelsonprimary@cengage.com

For learning solutions, visit **cengage.com.au**

Printed in China by 1010 Printing International Ltd
1 2 3 4 5 6 7 29 28 27 26 25

Nelson acknowledges the Traditional Owners and Custodians of the lands of all First Nations Peoples. We pay respect to Elders past and present, and extend that respect to all First Nations Peoples today.

Contents

Are There Lifeforms on Other Planets?	**4**
Possible Extraterrestrial Lifeforms	**10**
Positives and Negatives of Discovering Life Beyond Earth	**14**
The Search for Evidence	**17**
Extraterrestrial Visitors?	**22**
A Worthy Search	**30**
Glossary	**31**
Index	**32**

Are There Lifeforms on Other Planets?

People have long wondered if there might be life elsewhere in the universe, in our **galaxy** or beyond. For many hundreds of years, people have wondered if other planets like Earth exist, and if so, what the lifeforms on them might look like. They've also wondered what sort of relationship humans might have with these **extraterrestrial** lifeforms if we encountered them.

Extraterrestrial life could take many different forms. Most scientists think the most likely form of extraterrestrial life is **microbes**, such as **bacteria**. It could also take the form of plants or animals that have adapted to the conditions of their home planets. They might look very different from animals and plants on Earth.

Some people think we have already had visitors on Earth from other planets and claim to have seen these visitors with their own eyes.

Could life exist in another galaxy, such as the Small Magellanic Cloud galaxy?

For many decades, scientists have investigated whether life exists beyond Earth, using a range of methods. These include observing distant planets using powerful **radio telescopes** and making digital maps of galaxies. These maps and observations help scientists identify planets that could possibly have suitable conditions for life to arise. From studying how life has evolved on Earth, scientists have been able to identify many of the features a planet requires to be life-friendly, such as having liquid water on its surface and being neither too hot nor too cold.

Powerful radio telescopes help scientists scan the universe in search of life.

Early Imaginations

The Ancient Greeks may have been some of the earliest people to think about life on other planets. For example, the **philosopher** Epicurus (341–270 BCE) believed that there were many worlds other than our own, and that animals and plants were likely to exist in these worlds too.

Epicurus

Scientists who think life is unlikely to exist on other planets point out that planets need particular features to support life. For example, Earth contains liquid water, which is needed for all known lifeforms to survive. Earth also has an **atmosphere** that contains oxygen and is not too hot or cold. None of the other planets in our solar system share these conditions.

However, other scientists point out that there are trillions of stars and galaxies in the universe. Because of this, they think there may be many planets in the universe that could offer suitable conditions for life. In fact, scientists have already found several Earth-like planets in the Milky Way and other galaxies near it. Some of these planets contain water and other substances that could make life possible there. Still, no **evidence** of life on these planets has yet been found.

Earth has unique features that make it possible for life to evolve.

Habitable Zones

To support life, planets need to be in an area called a **habitable** zone. This is an area near a star within which liquid water can exist on a planet's surface if the planet has the right atmosphere. In our solar system, the star that creates a habitable zone is the Sun, and only Earth, Mars and Venus have atmospheres that are **hospitable** to life.

Mars and Venus are both near the edges of the habitable zone, but scientists think it's unlikely that they have liquid water on their surfaces. Venus, being close to the Sun, is very hot. Mars, further away from the Sun, is very cold. Both planets might be suitable for microbes, such as bacteria or **algae**, but so far no life has been found on either of them.

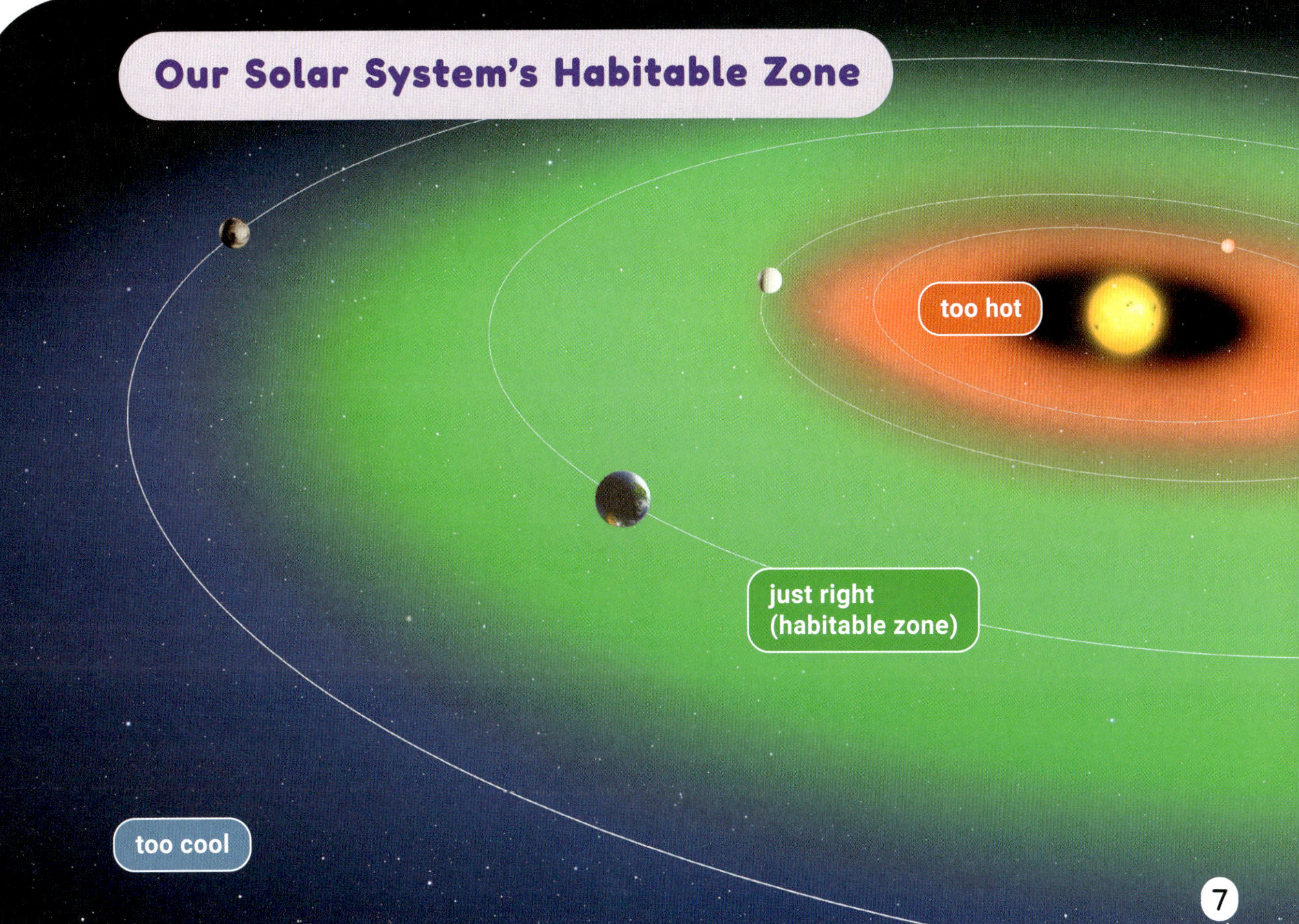

Gliese 12b is a recently discovered planet that has similar conditions to Earth.

Planets that **orbit** stars other than our own Sun are known as **exoplanets**. In our galaxy, the Milky Way, thousands of exoplanets have been discovered. As telescopes become more powerful, scientists expect to find more – possibly millions more. Many scientists think that life will eventually be found on some of the exoplanets they'll find in habitable zones.

The Milky Way Galaxy

Tens of Millions of Worlds

Scientists currently think the universe contains about 10 million trillion stars! Most stars have planets orbiting them, so there is an even higher number of planets in the universe.

Possible Extraterrestrial Lifeforms

Microbes and Bacteria

Extraterrestrial life could exist in the form of tiny microbes, or even plant-like life. These may prove hard to find, as microbes could be living in oceans below the surfaces of other planets.

But we may not have to look to other planets to find extraterrestrial microbes. Scientists working from the International Space Station have discovered that certain types of bacteria from Earth can survive in space. This makes some scientists think that bacteria might eventually be found floating in space.

Some bacteria such as these have been found to survive in space.

Animals

It's also possible that animal life exists on other planets. Extraterrestrial animals might have existed much longer on their planets than humans have existed on Earth. As a result, they might be more advanced than humans in particular ways. For example, they might have senses that humans don't have, or knowledge about things that humans are yet to discover. On the other hand, it is possible that extraterrestrial animals have existed for a shorter time and might not be as advanced as we are.

On other planets, it may be harder to distinguish plants from animals.

Humanlike Civilisations

It's also possible that we may one day find **civilisations** that are like our own in a range of ways. Intelligent lifeforms may have developed the abilities to communicate with each other and work together in ways that humans would recognise.

Carbon-Based Lifeforms

There may even be lifeforms that can survive in vastly different conditions from those on Earth. All life on Earth is based on a substance called carbon; carbon is the main "building block" of life. But lifeforms on other planets might have other substances as their main building blocks. This might allow them to survive in much hotter or colder conditions than those we experience on Earth.

There is a wide range of different lifeforms on Earth.

Sci-Fi Aliens

Since the 1960s, science-fiction movies have given many people the idea that extraterrestrial beings would look somewhat like humans, with bigger eyes and different body shapes. However, the aliens in science-fiction books in the 1800s were often more like big insects or sea creatures!

This illustration is from the 1901 book *The First Men in the Moon.*

Positives and Negatives of Discovering Life Beyond Earth

As interesting as it is to imagine what life on other planets might be like, all sorts of things could happen if we do find proof of extraterrestrial life. There could be both positive and negative outcomes from such a discovery.

Positives

Even if we only ever find extraterrestrial life in microbe or plant form, this could help us learn more about life on Earth and how it began. The knowledge we gain from studying these lifeforms could even help us develop new foods or medicines.

Extraterrestrial plants could have different qualities from the plant life on Earth.

If we were to discover intelligent extraterrestrial lifeforms, we might be able to learn a lot from them. For example, intelligent extraterrestrials might be much further advanced in technology than we are. They might be able to share their knowledge with us in areas such as maths, medicine, science or space travel.

If extraterrestrial beings are friendly, we could learn a great deal from them.

To visit Earth, extraterrestrial beings would need much faster spacecraft than the rockets currently used by Earth's astronauts. With our current spaceships, it would take us thousands of years to reach our closest neighbouring star, Proxima Centauri.

Negatives

Contact with extraterrestrial life might have negative consequences, too. These could include the introduction of diseases or poisonous substances that don't exist on Earth.

If humans ever interact with intelligent extraterrestrial lifeforms, other problems could arise. For example, we would probably have difficulties in communicating with them, which could lead to misunderstandings or conflict. They might not want to cooperate with humans, or might even want to make their own use of Earth's resources.

If extraterrestrial life arrived on Earth, we might not be able to understand what they want.

It's also possible that humans could do harm to the extraterrestrial lifeforms we encounter. For example, they might be **vulnerable** to diseases we could pass on to them.

Hopefully, however, contact with other lifeforms would benefit both us and them.

The Search for Evidence

A range of methods is used to search for life beyond Earth. These include listening for **radio signals** from other civilisations and searching with powerful telescopes and space vehicles. Computers are also used to create digital models that suggest where life might be found elsewhere in the universe.

Scientists at NASA meet to discuss the latest research on life beyond Earth.

Radio Signals (SETI)

On Earth, radio telescopes search for extraterrestrial life. These telescopes detect radio signals from space. Many radio telescopes work together in a search known as SETI, the Search for Extraterrestrial Intelligence. SETI researchers hope that advanced extraterrestrial civilisations use radio signals to communicate with each other, like humans do. They hope that if we keep listening, we'll eventually receive extraterrestrial communication, even though we probably won't understand what it is saying. So far, no extraterrestrial signals have been received.

Telescope Research

Telescopes play an important role in the search for planets in the habitable zones of other stars. The most powerful telescopes in the search are those that have been launched into space. One of the first space telescopes to join the search was **NASA**'s Kepler space telescope. Launched in 2009, Kepler discovered many possibly habitable planets in our galaxy.

The Kepler telescope scanned stars for the slight dimming of light that happens when an orbiting planet crosses the star's face.

The James Webb Space Telescope, launched in 2021, can study the atmospheres of exoplanets. It does this by capturing light from stars as it passes through the atmospheres of their exoplanets. Different gases in planets' atmospheres affect the light's brightness in different ways. Scientists can examine the differences in light to work out which gases are in a planet's atmosphere. This can tell them if a particular planet might have an atmosphere that is suitable for life.

This image of a young, star-forming region of the Carina Nebula was taken by the James Webb Space Telescope.

Life on Venus?

Some scientists think there may once have been life on Venus, in the form of microbes. Venus was once much cooler than it is now and may have had liquid water on its surface, making it more likely that life could have evolved there.

Computer Modelling

Scientists use computers to make digital models that show how stars, planets and galaxies are formed. They hope this will teach us more about the ideal conditions for life to form, and help us identify exoplanets that are suitable for life. For example, the European Space Agency's Gaia mission is currently using a space telescope to create a 3D map of the Milky Way. This map will help scientists closely study the location of stars and planets and identify the planets most likely to support life.

The Gaia satellite was launched in 2013 and is expected to operate until 2025.

Space Missions

Scientists also search for extraterrestrial life through space missions. NASA's Mars 2020 Perseverance **rover** has looked for signs of life while exploring the surface of Mars. The Perseverance rover has already found the remains of an ancient lake, which shows that Mars once held liquid water. Scientists hope that soil and rock samples collected by the rover may contain fossils of ancient life, though it may be many years before the samples can be sent to Earth for study.

Perseverance rover

Extraterrestrial Construction?

Scientific thinking about extraterrestrial life can change rapidly. For example, some scientists think that extraterrestrial lifeforms might have built huge structures around distant stars to harness their energy. In May 2024, a group of scientists identified seven stars that could possibly have these structures around them. However, in June that year other scientists released evidence that these stars may actually be galaxies surrounded by hot dust.

Extraterrestrial Visitors?

So far, no proof has been found that life exists on other planets. However, some people claim to have evidence that beings from other planets exist and have visited Earth.

Sometimes this evidence is in the form of photos or videos. Often, however, we only have the words of people who claim to have seen extraterrestrial spacecraft or interacted with extraterrestrial visitors.

Many people claim to have seen extraterrestrials in Nevada, USA, and now many tourists visit the area because of this.

Unknown objects seen in the sky are known as **UAPs (unidentified aerial phenomena)**. People who see UAPs often assume that they are seeing extraterrestrial spacecraft, but there are often simple explanations for unusual sightings. We can easily mistake natural **phenomena** such as comets or clouds or human-made objects such as aircraft or spy equipment for extraterrestrial lights or spacecrafts.

Nevertheless, over the centuries, so many sightings of extraterrestrial aircraft and lifeforms have been reported that some people think they can't all be mistaken. Some reports go back thousands of years. For example, in Ancient Rome, many sightings were recorded of shiny metallic vehicles in the sky.

This painting has made people wonder if extraterrestrial aircraft visited the Ancient Egyptian civilisation.

UFOs or UAPs?

UAPs used to be called unidentified flying objects (UFOs). The term UAP is broader, including all phenomena in the sky, not just objects.

Over the last century, many sightings of and encounters with UAPs have been reported. These usually haven't been backed up with evidence, so scientists often have strong doubts about them.

Most stories about UAPs turn out to be hoaxes.

Let's take a look at some of the most famous sightings of the last century.

The Roswell Case

Roswell, USA, 1947

Perhaps the most famous UAP case is a series of events that took place in 1947 near the city of Roswell in New Mexico, USA. At the time, many people were reporting seeing unidentified objects in the sky. When debris was found in the desert, many people believed it to be wreckage from a UAP crash. Some people even claimed to have found and filmed the bodies of extraterrestrial creatures that had died in the crash.

The US military explained that the wreckage was from a weather balloon or a spy balloon, but many people still believe that the debris came from an extraterrestrial spacecraft.

An intelligence officer recovers material from the Roswell crash site.

Summer of the Saucers

In mid-1947, so many saucer-shaped vehicles were reported in the skies over the USA that the summer of 1947 became known as "the summer of the saucers".

The Westall UAP Sighting

Melbourne, Australia, 1966

In 1966, in the Melbourne suburb of Clayton South, many people claimed to have seen a circular silver-coloured object hovering and landing in a patch of bushy land near Westall High School. Witnesses included teachers from the high school and many of its students. Some witnesses claimed to have seen two other similar craft, higher up in the sky.

Possible explanations were that the object was a weather balloon or a **radiation**-monitoring balloon that had blown over from South Australia. But the sightings have never been fully explained.

This photo is supposedly of the UAP seen at Westall in 1966.

The Kaikoura Lights

Aotearoa New Zealand, 1978

On 30 December 1978, many passengers onboard a plane flying over Aotearoa New Zealand's South Island observed strange lights from the aeroplane window. The lights seemed to be following the plane. The lights were detected on **radar**, and one passenger even filmed them.

Investigations by the New Zealand air force, police and an **observatory** in the city of Wellington suggested that the lights had come from boats and had reflected off clouds near the plane. The investigations also suggested that the lights may have come from the planet Venus, from vehicles on the ground, or from **meteors**.

The Kaikoura lights were unusual and mysterious but many explanations were suggested for them.

Rendlesham Forest Sightings

United Kingdom, 1980

In December 1980, in Rendlesham Forest in Suffolk, UK, there were reports of strange lights over the trees. Some witnesses claimed to have seen a large UAP with red and blue lights that hovered in the air and moved away from them when they approached it.

It is possible that the witnesses had simply seen lights from a nearby lighthouse, or an aircraft from a nearby military base.

This sculpture in the Rendelsham Forest recalls the sighting of a strange shape in the sky.

Iinomachi, Japan Present Day

The rural community of Iinomachi in Fukushima, Japan, is regarded as a current UAP hotspot. Since the 1980s, apparently hundreds of people in the area have reported seeing bright lights and saucer-shaped flying objects in the sky there, particularly near a cone-shaped mountain nearby called Senganmori. Locals take these reports so seriously that there is now a UAP research centre in Iinomachi.

Some researchers suggest that witnesses may simply have seen spy balloons from other countries. Others suggest that reports of sightings may have been exaggerated to attract tourists to the area.

There have been many reports of strange lights and objects in the sky in Iinomachi, Japan.

An Extraterrestrial Mountain?

Some residents of Iinomachi believe that Mount Senganmori was built by extraterrestrial creatures, and that these creatures have constructed a base underneath it.

A Worthy Search

As technology continues to develop, the likelihood that we will find life beyond Earth will increase. Even if proof is never found, the search for extraterrestrial life will continue to motivate scientists to learn more about the universe and our neighbouring galaxies. We may benefit in unexpected ways from the new technologies that are developed in the search.

More importantly, looking beyond Earth reminds us how unique and beautiful the planet we live on is, and can remind us to take better care of it.

Earth, viewed from the Moon

Glossary

algae (*noun*)	simple plants like seaweed that grow in or near water
atmosphere (*noun*)	a mixture of gases surrounding Earth or another planet
bacteria (*noun*)	tiny organisms, some types of which can cause disease
BCE (*adjective*)	Before the Common Era; the number of years before the time dates are counted from
civilisations (*noun*)	organised, advanced communities
evidence (*noun*)	information or facts that can help determine whether an idea is true or false
exoplanets (*noun*)	planets that orbit around stars that are not our Sun
extraterrestrial (*adjective*)	something that comes from beyond Earth or its atmosphere
galaxy (*noun*)	a huge group of stars and solar systems
habitable (*adjective*)	providing good conditions for living or growing
hospitable (*adjective*)	an area suitable to be lived in
meteors (*noun*)	a rock or other object that enters Earth's atmosphere from space
microbes (*noun*)	tiny living things such as bacteria
NASA (*noun*)	the National Aeronautics and Space Administration
observatory (*noun*)	a large building that houses a telescope
orbit (*verb*)	to circle another object in a regular pattern
phenomena (*noun*)	more than one phenomenon. A phenomenon is an event that can be observed or experienced in some way
philosopher (*noun*)	a deep thinker who seeks knowledge and wisdom
radar (*noun*)	a system that uses radio waves to detect the movement of vehicles such as ships or aeroplanes
radiation (*noun*)	a form of energy that can be harmful to living things
radio signals (*noun*)	signals that can carry radio broadcasts through the air or through space
radio telescopes (*noun*)	types of telescopes that detect radio signals
rover (*noun*)	a small vehicle that can explore the surface of a planet
UAPs (unidentified aerial phenomena) (*noun*)	objects seen in the sky that cannot be explained
vulnerable (*adjective*)	able to be harmed by something

Index

algae 7, 31
Ancient Egypt 23
Ancient Greece 5
Ancient Rome 23
animals 4, 5, 11
atmospheres 6, 7, 19, 31
bacteria 4, 7, 10, 31
balloons 25, 26, 29
carbon-based lifeforms 12
Carina Nebula 19
civilisations 12, 17, 21
computer modelling 20
Earth 4, 5, 6, 7, 8, 10, 11, 12, 14-16, 17, 21, 22, 30, 31
Epicurus 5
European Space Agency 20
evidence 6, 17-21, 22, 24, 31
exoplanets 8, 17, 20, 31
extraterrestrial construction 21
extraterrestrial life 4, 10-12, 14-16, 22-24, 25, 26, 27, 28, 29, 30, 31
fossils 20
Gaia satellite 20
galaxies 4, 6, 8-9, 18, 20, 21, 30, 31
Gliese 12b 8
habitable zones 7, 8, 18, 31
Iinomachi, Japan 29
International Space Station 10
James Webb Space Telescope 19
Kaikoura Lights, Aotearoa New Zealand 27
Kepler space telescope 18
Mars 7, 21
Mars rover 21
meteors 27, 31
microbes 4, 7, 10, 14, 19, 31
Milky Way 6, 8-9, 20
Mount Senganmori 29
NASA 18, 21, 31
observatories 27, 31
orbits 8, 31
Perseverance rover 21
phenomena 23, 31
philosophers 5, 31
planets 4-8, 9, 10, 11, 12, 14, 18, 19, 20, 22, 27, 31
plants 4, 5, 10, 11, 14
Proxima Centauri 15
radars 27, 31
radiation 26
radio signals 17
radio telescopes 5, 17
Rendelsham Forest, UK 28
Roswell, USA 25
rovers 21, 32
sci-fi aliens 13
SETI, the Search for Extraterrestrial Intelligence 17
Small Magellanic Cloud galaxy 4
solar system 6, 7, 31
space missions 21
spacecraft 15, 22, 23
stars 6, 7, 8, 9, 15, 18, 19, 20, 21, 31
Sun 7, 8, 9
The First Men in the Moon 13
UAPs (unidentified aerial phenomena) 23-24, 25, 26, 27, 28, 29
UFOs 23
Venus 7, 19, 27
vulnerability 16
water 5, 6, 7, 10, 19, 21, 31
Westall, Melbourne, Australia 26